GOODS OF THE MIND, LLC

Competitive Mathematics Series

for

Gifted Students in Grades 3 and 4

PRACTICE ARITHMETIC

Cleo Borac, M. Sc.
Silviu Borac, Ph. D.

This edition published in 2014 in the United States of America.

Editing and proofreading: David Borac, M.Mus.
Technical support: Andrei T. Borac, B.A., PBK

Send all inquiries to:

Goods of the Mind, LLC
1138 Grand Teton Dr.
Pacifica
CA, 94044

Competitive Mathematics Series for Gifted Students
Level II (Grades 3 and 4)
Practice Arithmetic
2nd edition

Contents

CONTENTS

FOREWORD

The goal of these booklets is to provide a problem solving training ground starting from the earliest years of a student's mathematical development.

In our experience, we have found that teaching how to solve problems should focus not only on finding correct answers but also on finding better solution strategies. While the correct answer to a problem can typically be obtained in several different ways, not all these ways are equally useful for learning how to solve problems.

The most basic strategy is *brute force*. For example, if a problem asks for the number of ways Lila and Dina can sit on a bench, it is easy to write down all the possibilities: Dina, Lila and Lila, Dina. We arrive at this solution by performing all the possible actions allowed by the problem, leaving nothing to the imagination. For this last reason, this approach is called brute force.

Obviously, if we had to figure out the number of ways 30 people could stand in a line, then brute force would not be as practical, as it would take a prohibitively long time to apply.

Using brute force to obtain the correct answer for a simpler problem is not necessarily a useful learning experience for solving a similar problem that is more complex. Moreover, solving problems in a quantitative manner, assuming that the student can transfer simple strategies to similar but more complex problems, is not an efficient way of learning problem solving.

From this simple example, we see that the goal of *practicing* problem solving is different from the goal of problem solving. While the goal of problem solving is to obtain a correct answer, the goal of practicing problem solving is to acquire the ability to develop strategies, generate ideas, and combine approaches that are powerful enough to solve the problem at hand as well as future similar problems.

While brute force is not a useless strategy, it is not a key that opens every door. Nevertheless, there are problems where brute force can be a useful tool. For instance, brute force can be used as a first step in solving a complex problem: a smaller scale example can be approached using brute force to help the problem solver understand the mechanics of the problem and generate ideas for solving the larger case.

All too often, we encounter students who can quickly solve simple problems by applying brute force and who become frustrated when the solving methods they have been employing successfully for years become inefficient once problems increase in complexity. Often, neither the student nor the parent has a clear understanding of why the student has stagnated at a certain level. When the only arrows in the quiver are guess-and-check and brute force, the ability to take down larger game is limited.

Our series of books aims to address this tendency to continue on the beaten path - which usually generates so much praise for the gifted student in the early years of schooling - by offering a challenging set of questions meant to build up an understanding of the problem solving process. Solving problems should never be easy! To be useful, to represent actual training, problem solving should be challenging. There should always be a sense of difficulty, otherwise there is no elation upon finding the solution.

Indeed, practicing problem solving is important and useful only as a means of learning how to develop better strategies. We must constantly learn and invent new strategies while questioning the limitations of the strategies we are using. Obtaining the correct answer is only the natural outcome of having applied a strategy that worked for a particular problem in the time available to solve it. Obtaining the wrong answer is not necessarily a bad outcome; it provides insight into the fallacies of the method used or into the errors of execution that may have occured. As long as students manifest an interest in figuring out strategies, the process of problem solving should be rewarding in itself.

Sitting and thinking in a focused manner is difficult to train, particularly since the modern lifestyle is not conducive to adopting open-ended activities. This is why we would like to encourage parents to pull back from a quantitative approach to mathematical education based on repetition, number of completed pages, and the number of correct answers. Instead, open up the time boundaries that are dedicated to math, adopt math as a game played in the family, initiate a math dialogue, and let the student take his or her time to think up clever solutions.

Figuring out strategies is much more of a game than the mechanical repetition of stepwise problem solving recipes that textbooks so profusely provide, in order to "make math easy." Mathematics is not meant to be easy; it is meant to be interesting.

Solving a problem in different ways is a good way of comparing the merits of each method - another reason for not making the correct answer the primary goal of the activity. Which method is more labor intensive, takes more time or is more prone to execution errors? These are questions that must be part of the problem solving process.

In the end, it is not the quantity of problems solved, the level of theory absorbed, or the number of solutions offered in ready-made form by so many courses and camps, but the willingness to ask questions, understand and explore limitations, and derive new information from scratch, that are the cornerstones of a sound training for problem solvers.

These booklets are not a complete guide to the problem solving universe, but they are meant to help parents and educators work in the direction that, aside from being the most efficient, is the more interesting and rewarding one.

The series is designed for mathematically gifted students. Each book addresses an age range as some students will be ready for this content earlier, others later. If a topic seems too difficult, simply try it again in a couple of months.

DIGITS, NUMBERS AND CRYPTARITHMS

Numbers are written using *digits*. In decimal notation, there are 10 available digits: 0, 1, 2, 3, 4, 5, 6, 7, 8, and 9.

An integer can be expressed in *expanded form* as follows, so that the digits and their place values are separate:

$$5555 = 5 \times 1000 + 5 \times 100 + 5 \times 10 + 5$$

addition

The *digit sum* of a whole number is the sum of its digits. The digit sum is denoted with S:

$$S(12) = 1 + 2 = 3$$
$$S(452) = 4 + 5 + 2 = 11$$

$S(23) = 5$

Facts used in problem solving:

- The digit sum of a two digit number is at most 18.
- The digit sum of a three digit number is at most 27.
- Adding digits of zero to a number does not change its digit sum.
- Removing digits of zero from a number does not change its digit sum.
- The digit sum of a non-zero integer cannot be zero.

multiplication

The *digit product* of a whole number is the product of its digits. The digit product is denoted with P:

$$P(12) = 1 \times 2 = 2$$
$$P(452) = 4 \times 5 \times 2 = 40$$

Facts used in problem solving:

- Adding digits of 1 to a number does not change its digit product.
- Removing digits of 1 from a number does not change its digit product.
- An even number cannot have an odd digit product.
- A number with a digit product of 1 is a repunit.

Cryptarithms are mathematical riddles. A cryptarithm consists of a simple operation, such as an addition or a multiplication, in which some or all the digits are replaced by symbols (*encrypted*).

The rules of the cryptarithm are often part of the problem statement. Rules for cryptarithms are generally, but not always, as follows:

- different symbols represent different digits,
- the same symbol represents the same digit.
- no number starts with a digit of zero.

A *repdigit* is a number in which all the digits are identical. Examples of repdigits:

$$5555555$$
$$33333$$
$$\underbrace{77\cdots7}_{1980\ \text{times}}$$

A *repunit* is a special case of repdigit in which all the digits are equal to 1.

Some properties of repdigits that are useful in problem solving are based on the following fact:

$$999999 = 1000000 - 1$$

Another example:

$$\underbrace{99\cdots99}_{3219\ \text{times}} = 1\underbrace{00\cdots00}_{3219\ \text{times}} - 1$$

PRACTICE ONE

> Do not use a calculator for any of the problems!

Exercise 1

[handwritten: 10 digits]

1. The largest integer with all different even digits is ....86420.... ✓
2. The largest integer with all different odd digits is 9.75.3.1. ✓
3. The largest integer with all different digits is 98.7.6.5.4.3.2.1.0
4. The next integer with all different digits after 69785 is ...4.3.1. 69801
5. The largest 6-digit integer with a digit product of 12 is ...?.... 621111
6. The largest palindrome written with three 5s and six 3s is .33.3.5.5.5.3.3.3 *[handwritten: 33 355 53 3 3]*
7. The largest 6 digit integer with a digit sum of 12 is .9.3.0000. 533353335
8. The smallest 6 digit integer with a digit sum of 12 is .100029...

Exercise 2

All the digits of a number A are different. A number can be formed using digits that are each larger than the digits of A by one. What is the largest such number A? *[handwritten: 876543210]*

Exercise 3

All the digits of a number B are different. A number can be formed using digits that are the double of the digits of B. What is the largest such number B?

11

Exercise 4

Stephan, the tennis coach, wrote an invoice for one of his clients. He noticed that the number of the invoice was 91087 and asked himself: "After how many more invoices will there be another invoice number in which all the digits are different?"

Exercise 5

1. How many digits are multiples of 3?
2. How many digits are multiples of 4?
3. How many digits are multiples of 5?
4. How many digits are multiples of 7?

Exercise 6

The digits m, n, and s can be sorted in increasing order like this: $m > n > s$. With these digits, we can form 6 three-digit numbers. Which of the following cryptarithms are true? Check all that apply.

(**A**) $msn > nms > smn$

(**B**) $nms > mns > smn$

(**C**) $mns > nsm > snm$

(**D**) $nms > nsm > msn$

Exercise 7

The number 44444 can be written as the sum of two integers that differ by 6666. Find the value of the smaller integer.

Exercise 8

If the tens digit of a number is 4 times the units digit and the hundreds digit is twice the tens digit, is the number even or odd?

Exercise 9

Some numbers have a digit product of 30 and, when multiplied by 5, produce a number that ends in zero. What is the largest 6-digit number of this kind?

Exercise 10

What are the first three digits of the product: 999999999999×6?

Exercise 11

Three consecutive numbers have digit sums of 18, 1, and 2. What is the digit sum of their sum?

Exercise 12

How many positive integers with non-zero digit products have digit sums of 1, 2, or 3?

Exercise 13

What is the smallest positive integer with a digit sum of 20?

Exercise 14

How many digits does the smallest positive integer with a digit sum of 203 have?

Exercise 15

The smallest positive integer with a digit sum of 1003 has:

(A) 101 occurences of the digit 9

(B) 110 occurences of the digit 9

(C) 111 occurences of the digit 9

(D) no digits of 9

Exercise 16

What is the remainder when we divide each of the following by 3?

$$10, \ 100, \ 1000, \ 10000, \ 100000$$

Exercise 17

What is the remainder when we divide each of the following by 9?

10, 100, 1000, 10000, 100000

Exercise 18

What is the remainder when we divide each of the following by 11?

10, 100, 1000, 10000, 100000

SEQUENCES

Add or subtract a fixed number from a term to generate the next term of an **arithmetic sequence**:

$$\overset{+4 \quad +4 \quad +4}{5, 9, 13, 17, 21, \ldots}$$

Multiply or divide a term by a fixed number to generate the next term of a **geometric sequence**:

$$\overset{\times 3 \quad \times 3 \quad \times 3}{6, 18, 54, 172, \ldots}$$

Example: Identify the sequences that are arithmetic and the ones that are geometric:

1. $0, 3, 6, 9, \cdots$
2. $1, 3, 9, 27, \cdots$
3. $1, -3, 9, -27, \cdots$
4. $4, 3, 2, 1, 0, \cdots$
5. $10, 100, 1000, 10000, \cdots$

Experiment

Write some arithmetic and geometric sequences of your choice.

There are also a large variety of ad-hoc sequences. A few examples:

1. a sequence that alternates terms from two different arithmetic or geometric sequences:

$$2, \mathbf{10}, 4, \mathbf{7}, 6, \mathbf{4}, 8, \mathbf{1}, 10, \mathbf{-2}, \cdots$$

2. an apparently random sequence:

$$11, 44, 3, -9, 103, 7, 8, 41, \cdots$$

3. a sequence formed by the digit sums of consecutive numbers:

$$1, 2, 3, 4, 5, 6, 7, 8, 9, 1, 2, 3, 4, 5, 6, \cdots$$

4. a sequence formed by the remainders from dividing consecutive numbers by 4:

$$1, 2, 3, 0, 1, 2, 3, 0, \cdots$$

Experiment

Create a few sequences using rules of your own design.

Often, problems propose more complex rules for generating sequences. Experiment with these sequences to find out more about them!

Experiment

Write the first 10 terms of a sequence with a first term of 5, in which each subsequent term is generated by the rules:

- multiply the current term by 2,
- take the digit sum of the result.

What do you notice?

Experiment

Write the first 10 terms of a sequence with a first term of 0 in which each subsequent term is generated by the rules:

1. add 5 to the current term,
2. multiply the result by 2,
3. take the quotient when dividing by 3.

What do you notice?

Experiment

Try to make up sets of rules yourself and generate a sequence using them. What do you notice?

The **Fibonacci sequence** is a sequence that starts with two terms of 1, where each subsequent term is formed by adding the two previous terms:

$$1, \ 1, \ 2, \ 3, \ 5, \ 8, \ 11, \ \cdots$$

Experiment

Write a few more terms to the Fibonacci sequence above.

The position of a term in a sequence is called **rank**. In the following sequence, the term 4 has rank 3 and the term 6 has rank 4:

$$1, \ 2, \ 4, \ 6, \ 8, \ \cdots$$

Example:

In the sequence:

$$1, \ -2, \ 3, \ -4, \ 5, \ -6, \ , \ \cdots$$

1. Which term has rank 4?
2. What is the rank of the term -8?
3. What is the rank of the largest term smaller than 15?
4. Which term has rank 8?
5. What is the rank of the smallest term larger than -10?
6. Is there a term that is the same as its rank?

Answers:

1. The term -4 has rank 4.
2. The rank of the term -8 is 8.
3. The largest term smaller than 15 is 13. The rank of the term 13 is also 13 (it is the 13th term in the sequence).
4. The term -8 has rank 8.
5. The smallest term larger than -10 is -8 and has rank 8 (it is the 8th term in the sequence).
6. All odd terms are the same as their ranks.

PRACTICE TWO

Do not use a calculator for any of the problems!

Exercise 1

Write three more terms to the following sequences:

    1. 10, 11, 100, 101, 110, 111, 1000, $\cdots$

    2. 3, 4, 3, 4, 5, 6, 5, 6, 7, $\cdots$

    3. 1, 2, 4, 7, 11, $\cdots$

    4. 11, 22, 66, 264, $\cdots$

Exercise 2

Lila divided the numbers in the following list by 10 and placed the remainders in boxes, one number per box. How many boxes did she use?

    71, 99, 32, 401, 63, 27, 49, 111, 60, 94, 85, 868, 18, 19

Exercise 3

Dina had to divide positive consecutive integers starting from 1 by 7 and write down the sequence formed by the remainders. Soon, Dina realized she no longer needed to write more of the sequence: she was able to predict any future term. For example, Dina could predict the 45th term in the sequence. Which one of the following numbers matches Dina's prediction?

(A) 2

(B) 3

(C) 4

(D) 6

Exercise 4

What is the remainder when we divide the following number, in which some digits have been hidden by symbols, by 100?

$$7\spadesuit\diamondsuit 8\clubsuit 3\diamondsuit\heartsuit 66\diamondsuit 2\clubsuit\clubsuit\heartsuit 532$$

Exercise 5

Amira and Dina used some chalk to create a new game in the driveway. They drew 6 squares in a line and marked them with the letters A, B, C, D, M, and N. Dina started at A and jumped to B, then to C, and so on until square N. On square N, she turned around and jumped from square to square in the opposite direction. The same thing happened when she reached square A again, and so on. Which letter was on the square Dina landed on after her 28th jump?

Exercise 6

Three bunnies, Keef, Leef, and Meef, stand in line with Keef first and Meef last. They start jumping, repeating the following rules:

- the last jumps over the middle one,
- the middle one jumps over the first one.

After how many jumps will Meef be last in line again?

Exercise 7

Dina says "42" and Lila replies "6." Dina says "901" and Lila replies "10." Dina says "3" and Lila replies "3." Dina says "181" and Amira must guess Lila's reply. Help Amira choose the correct answer from the following:

(A) 18

(B) 10

(C) 5

(D) 1

Exercise 8

A machine takes a number as an input and outputs a number related to it. This is a list of inputs and outputs:

input	151	78	100	91
output	5	56	0	9

What will the machine output if the input is 345?

Exercise 9

What is the next term in the sequence?

$$1, \ 2, \ 2, \ 4, \ 8, \ 32, \ \cdots$$

Exercise 10

If Dina guesses the next term of each of the following sequences correctly, she gets as many points as the term she guessed. She starts the game with zero points.

1. $Z, \ 1, \ Y, \ 2, \ X, \ 3, \ W, \ \cdots$

2. $1, \ 1, \ 2, \ 3, \ 5, \ 8, \ \cdots$

3. $1, \ 7, \ 5, \ 11, \ 9, \ 15, \ \cdots$

4. 7, 9, 3, 1, 7, 9, 3, $\cdots$

If Dina scored 18 points, how many times did she guess correctly?

(A) 1

(B) 2

(C) 3

(D) 4

Exercise 11

A sequence is generated by the following machines:

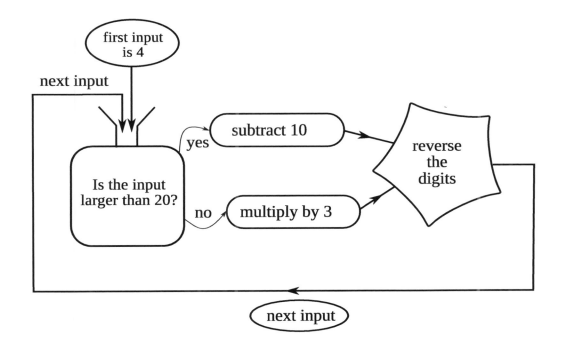

The first term is 4. After this first term is input, the machines keep generating more terms, feeding them in turn to the input.

Calculate the first 10 inputs.

Exercise 12

Dina cut a square into four small squares. Then, she cut one of the small squares into four squares. She repeated this operation 10 times, each time cutting into four one of the smallest squares available at the time. How many squares did she end up with?

THE SUM OF CONSECUTIVE INTEGERS

There are several ways of adding consecutive integers. In this age group, the most popular is to add the end numbers together and work our way inwards, like this:

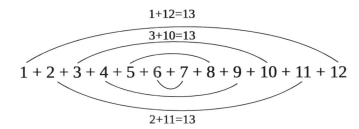

In the example shown, numbers can be paired to form 6 groups with a sum of 13. It is faster to multiply ($6 \times 13 = 78$) than to perform all the additions. This procedure, however, works smoothly only when the number of terms is even. A sum with an odd number of terms requires a little more work:

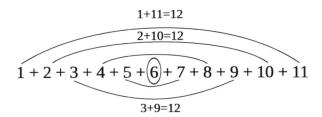

One must determine how many pairs of numbers with the same sum (12) can be formed and which number must be added separately since it has no pair.

A smoother method is to add each number twice, thereby making the total number of terms even. Add the pairs vertically, as follows:

$$\begin{array}{c} 1 + \ \ 2 + 3 + 4 + 5 + 6 + 7 + 8 + 9 + 10 + 11 \\ \underline{11 + 10 + 9 + 8 + 7 + 6 + 5 + 4 + 3 + \ \ 2 + \ \ 1} \\ 12 + 12 + 12 + 12 + 12 + 12 + 12 + 12 + 12 + \ 12 + 12 \end{array} +$$

There are 11 terms, each equal to 12. The sum we have to calculate, however, is only half of this, since we have added each term twice by design. The final result is:

$$1 + 2 + 3 + \cdots + 11 = 11 \times 12 \div 2 = 66$$

This last method helps us formulate a rule for computing the sum: *multiply the last term by the next consecutive integer and divide by 2.* This works, however, only if the first term of the sum is 1.

The result of such a sum is called a **triangular number**. Here is a list of small triangular numbers:

$$\begin{array}{rcl} 1 + 2 & = & 3 \\ 1 + 2 + 3 & = & 6 \\ 1 + 2 + 3 + 4 & = & 10 \\ 1 + 2 + 3 + 4 + 5 & = & 15 \\ 1 + 2 + 3 + 4 + 5 + 6 & = & 21 \end{array}$$

Triangular numbers result from multiplying two consecutive positive integers and dividing the product by 2.

Sums of consecutive numbers with an odd number of terms

Let us arrange the terms of the sum in increasing order. Notice that there is a middle number. Numbers that are equally far from the left and from the right sides of the middle number add up to twice the middle number:

$$57+59 = 58 + 58$$

$$56 + 57 + 58 + 59 + 60$$

$$56 + 60 = 58 + 58$$

In the example above, the sum of five consecutive numbers is five times as large as the middle number:

$$56 + 57 + 58 + 59 + 60 = 5 \times 58 = 290$$

Examples:

1. The sum of 7 consecutive numbers is 77. What is the middle number?

2. The sum of an odd number of consecutive numbers is 77. The middle number is 11. How many terms does the sum have?

3. An odd number (more than a single number) of consecutive numbers has a sum of 121. What is the middle number? How many terms does the sum have?

Answers:

1. The middle number is 11.

2. The sum has 7 terms.

3. The middle number is 11 and the sum has 11 terms.

PRACTICE THREE

Exercise 1

Compute the sums using the methods explained in chapter 6:

1. $1 + 2 + 3 + 4 + 5 + 6 + 7 =$
2. $1 + 2 + 3 + 4 + \cdots + 10 =$
3. $1 + 2 + 3 + 4 + \cdots + 50 =$
4. $1 + 2 + 3 + 4 + \cdots + 97 =$

Exercise 2

The sums below are not entirely similar to the ones we derived a formula for. Make the necessary adjustments to compute each of them:

1. $3 + 4 + 5 + 6 =$
2. $5 + 6 + 7 + 8 + 9 + 10 + 11 + 12 + 13 + 14 + 15 =$
3. $3 + 4 + 5 + \cdots + 99 =$
4. $10 + 11 + 12 + \cdots + 91 =$
5. $101 + 102 + 103 + \cdots + 150 =$

Exercise 3

The sums below are not entirely similar to the ones we derived a formula for. Make the necessary adjustments to compute each of them:

1. $2 + 4 + 6 + 8 + 10 =$
2. $2 + 4 + 6 + \cdots + 20 =$
3. $20 + 40 + 60 + \cdots + 200 =$
4. $10 + 12 + 14 + \cdots + 20 =$
5. $100 + 102 + 104 + \cdots + 110 =$

Exercise 4

The sums below are not entirely similar to the ones we derived a formula for. Make the necessary adjustments to compute each of them:

1. $10 + 20 + 30 + \cdots + 100 =$
2. $50 + 60 + 70 + \cdots + 100 =$
3. $1 + 3 + 5 + 7 + 9 + 11 + 13 =$
4. $101 + 103 + 105 + \cdots + 111 =$
5. $3 + 6 + 9 + 12 + \cdots + 36$

Exercise 5

A list of consecutive integers has a sum of 4. How many integers are there in this list?

Exercise 6

A list of consecutive integers has a sum of 11. How many integers are there in this list? Is there more than one answer?

Exercise 7

Mira is a dog breeder. Each of her Afghan hounds was born in a different consecutive year. When the sum of the ages of the hounds was 9, how old was the oldest dog? Is there more than one solution?

Exercise 8

The sum of three consecutive numbers is 31038. What is the middle number?

Exercise 9

Write the number 18105 as a sum of 6 consecutive multiples of 5.

Exercise 10

A fisherman's boat capsized in a storm and its catch of fish spilled onto a remote beach. A penguin came by and started eating fish at a rate of one fish every 2 minutes. As he was starting the second fish, another penguin arrived and started eating fish at the same rate. A new penguin arrived and began to eat fish each time the previous penguin started its second fish. How long did it take the penguins to eat the first 21 fish?

Exercise 11

How many digits does the following number have? Assume the pattern is the same to the end.

$$1223334444 \cdots 99$$

Exercise 12

Dina was completing a worksheet and had to compute the sum of 87 consecutive numbers. Arbax, the Dalmatian, played with the sheet and managed to damage it so that only the middle term of the sum was still legible: 100. What was the largest term of the sum?

Exercise 13

How many pairs of digits (a, b), with $a > b$, are there that satisfy the following cryptarithm? (The same letter represents the same digit. Different letters represent different digits.)

$$aaa + bbb = ccc$$

Exercise 14

Compute the sums:

1. $3 + 6 + 9 + \cdots + 99 =$
2. $4 + 8 + 12 + \cdots + 100 =$
3. $5 + 10 + 15 + \cdots + 100 =$
4. $6 + 12 + 18 + \cdots + 180 =$

Exercise 15

What is the value of k?

$$1 + 2 + 3 + \cdots + k = 820$$

DIVISIBILITY

Positive integers can be written in **expanded form** so that the digits and the place values are clearly separated:

$$984 = 9 \times 100 + 8 \times 10 + 4$$

Some divisibility rules:

Positive integers that have an even last digit are divisible by 2.

A positive integer is divisible by 4 if its last two digits form an integer that is divisible by 4.

A positive integer is divisible by 8 if its last three digits form an integer that is divisible by 8.

A positive integer is divisible by 5 if its last digit is divisible by 5 (it is either 5 or 0).

A positive integer is divisible by 3 if its digit sum is divisible by 3.

A positive integer is divisible by 9 if its digit sum is divisible by 9.

A positive integer is divisible by 11 if, starting from the rightmost digit, the sum of the digits in even places and the sum of the digits in odd places have a difference divisible by 11.

A positive integer is divisible by 10 if its last digit is 0.

The **digit sum** of a number is denoted by S. This is how we write the digit sums of 715168:

$$S(715168) = 7 + 1 + 5 + 1 + 6 + 8 = 28$$

To find out if the digit sum of a number is divisible by 3 or by 9, we can apply the digit sum test again:

$$S(28) = 2 + 8 = 10$$

Since 10 is not a multiple of 3, we can conclude that 715168 is not a multiple of 3. If a number is not a multiple of 3, it cannot be a multiple of 9 either.

If a number is divisible by 11, its **alternating digit sum** will be a multiple of 11 (including zero). For example:

$$\begin{aligned} A(87564312) &= 2 - 1 + 3 - 4 + 6 - 5 + 7 - 8 \\ &= 1 - 1 + 1 - 1 = 0 \end{aligned}$$

Therefore, the number 87564312 is a multiple of 11.

Any positive integer has a **set of divisors**, i.e. numbers it is divisible by. For example, this is the set of divisors of 100: 1, 2, 4, 5, 10, 20, 25, 50, and 100.

Practice Four

Exercise 1

How many of the following numbers are divisible by 4?

138, 142, 118, 442, 98, 146, 886, 2014, 76

Exercise 2

How many of the following numbers are divisible by 3?

96, 144, 32, 87, 0, 100, 51, 37, 57, 11

Exercise 3

Circle the numbers that are divisible by 3 but not divisible by 9.

15, 45, 32, 93, 111, 38, 0, 42, 108, 72, 69, 174, 153

Exercise 4

Find the largest number of the form 2♠8 that is divisible by 4 but not by 3.

Exercise 5

Circle the numbers that are not divisible by 11:

141, 101, 1001, 154, 1067, 111, 187, 209, 134

Exercise 6

Which of the following repunits are divisible by 11?

11, 111, 1111, 11111, 111111, 1111111

Fill in the blanks:

Repunits with .... *number of digits are divisible by* 11.

Does the rule you just found hold for any repdigit or only for repunits?

Exercise 7

Which of the following repdigits is divisible by 11?

1. $1 \underbrace{11 \cdots 1}_{111 \text{ digits}} 1$

2. $4 \underbrace{44 \cdots 4}_{444 \text{ digits}} 4$

3. $7 \underbrace{77 \cdots 7}_{777 \text{ digits}} 7$

4. $6 \underbrace{66 \cdots 6}_{666 \text{ digits}} 6$

Exercise 8

Dina, Lila, and Amira played a game in which they received a number with hidden digits. They had to find the hidden digits so that the number was divisible by both 2 and 3. Each of them was then allowed to add the numbers they found to their score. If the number they started with was 25♣, which of the following statements is true?

(A) Dina won because she got 252, the highest possible score.

(B) Lila won because she got 258, the highest possible score.

(C) Amira won because she got 510, the highest possible score.

Exercise 9

Alfonso, the grocer, has a list of all the products he received from a delivery van and wants to match this list to the order he placed. Unfortunately, the driver of the van spilled some coffee on the list and some of the digits are no longer legible:

Potatoes	3 ● 4	lbs
Lemons	1 6 0	lbs
Beets	● 5 9	lbs

Potatoes come in bags of 3 lbs each, lemons come in bags of 2 lbs each, and beets come in bags of 7 lbs each. Moreover, Alfonso is sure he ordered a smaller quantity of beets than of potatoes. Which of the following cannot be the total weight of the merchandise Alfonso received?

(A) 743 lbs

(B) 773 lbs

(C) 803 lbs

(D) 809 lbs

Exercise 10

Arbax, the Dalmatian, must guess the hidden digits for each number so that it is divisible by 3. For each number, Arbax may give one possible answer. For each correct answer, Arbax gets as many bones as the number he guessed. What is the largest number of bones Arbax can win?

1. 6♢4
2. ♣77
3. 11♠
4. 8♡1
5. 22♢

Exercise 11

Dina had to find all the numbers of the form $a55b$, where a and b are unknown digits, that are divisible by 15. How many such numbers are there?

Exercise 12

Alfonso received a shipment of onions. He packaged an equal number of onions in each bag and had some onions left over at the end. The sum of the number of onions in one bag and the number of bags was 21. The sum of the number of bags and the number of onions left over was 19. The sum of the number of onions in one bag and the number of onions left over was 10. How many onions were there in the shipment?

PRIMES

Use the following method to factor a number into primes:

1. Set the first prime divisor to be 2, the smallest prime number.

2. Divide the number by the prime divisor.

3. If the remainder is zero and the quotient is prime, stop. The factors derived so far form the complete prime factorization.

4. If the quotient is not prime and the remainder is zero, repeat this procedure on the quotient, starting from step 2.

5. If the remainder is not zero, set the divisor to be the next larger prime.

6. Apply this procedure to the quotient, starting with step 2.

This procedure has many practical advantages over factor trees, the method of choice in schools.

Example: Factor 1932 into primes.

1932	2
966	2
483	3
161	7
23	23
1	

1932 = 2 x 2 x 3 x 7 x 23

Facts about prime numbers used in problem solving:

1. 1 is not a prime number. (It is an *improper prime.*)
2. 2 is the only even prime.
3. A number has a single prime factorization.
4. The only way to tell if a number is prime is by factoring it.
5. 2 and 3 are the only primes that differ by 1.

Experiment

1. Which integer does M represent?

$$M \times 5 \times 11 = 11 \times 5 \times 17$$

2. Write 6 as a sum of primes.
3. Is 221 prime?

Check your answers:

1. Since the prime factorization of a number is unique, M must be equal to 17.
2. $6 = 2 + 2 + 2$. Of course, $6 = 1 + 2 + 3$ is not a correct answer because 1 is not prime.
3. No. $221 = 13 \times 17$. Remember you have to try to divide by all the primes smaller than the square root of 221. Since 225 (15×15) is the closest perfect square, 221 should be divided by all primes up to and including 13. 221 is not prime.

How do we decide that a number is prime? A number is prime if it cannot be divided exactly by any prime that ranges from 2 to the prime that is closest to the square root of the number.

If a number is the product of two primes, like this:

$$253 = 11 \times 23$$

we automatically obtain the factor 23 when we divide by 11. Since we try our divisions with ever increasing primes, there is no need to try to divide with primes larger than the "middle" one. When dealing with multiplication, the "middle" of a number is its square root.

Experiment

If we want to find out if 493 is prime, what is the range of primes we would try to divide it by?

1. 400 is 20×20. Therefore, the square root of 493 must be close to 20.

2. The primes closest to 20 are 19 and 23. 19 is surely in the range. Multiply 23 by 23 to see that the product is larger than the square root: $23 \times 23 = 529$.

3. We have to divide with primes ranging from 2 to 19.

4. 493 is not even, skip over division by 2.

5. $S(493) = 4 + 9 + 3 = 16$, skip over division by 3.

6. 493 does not end in 5 or 0, skip over division by 5.

7. $A(493) = 3 - 9 + 4 = -2$, skip over division by 11.

8. Attempt divisions by: 7, 13, 17, and 19

PRACTICE FIVE

Do not use a calculator for any of the problems!

Exercise 1

Factor into primes using the method explained in chapter 10:

1. 105
2. 42
3. 1331
4. 168
5. 504
6. 352

Exercise 2

Make a list of the primes you would use to factor the following numbers before deciding that they are prime. If you can use easy divisibility rules to rule out a prime divisor, do not put that prime on the list. For example, if the number is not even, do not put the divisor 2 on the list.

Beside each number, write the list of primes you tried and compare it with the solutions:

1. 151
2. 127
3. 191

   4. 257

   5. 233

   6. 331

   7. 499

Exercise 3

Factor the following into primes:

   1. 493

   2. 499

   3. 593

   4. 323

   5. 1007

   6. 731

   7. 481

Exercise 4

How many 3-digit prime numbers can be formed using each of the digits 1, 3, and 5 once?

Exercise 5

Amira wanted to find a 3-digit prime number of the form *aaa*. Is this possible? If so, which primes might she have found?

Exercise 6

Write 27 as a sum of distinct primes.

Exercise 7

Two prime numbers have a sum of 999. Which numbers are they?

Exercise 8

The sum of the first 999 prime numbers is:

(**A**) odd

(**B**) even

(**C**) neither

Exercise 9

Find three prime numbers that add up to another prime number.

Exercise 10

Find the smallest prime number that is the average of two prime numbers.

MISCELLANEOUS PRACTICE

Do not use a calculator for any of the problems!

Exercise 1

Dina has to replace X and Y by two different digits so that the following addition is correct.

$$\begin{array}{r} X\,Y\,X \\ X\,Y\,Y \\ \hline 1\,9\,7\,7 \end{array} +$$

Find the product of the digits X and Y.

Exercise 2

A positive integer is 4 times as large as another integer. Their difference cannot be:

(A) 51

(B) 513

(C) 5139

(D) 5191

Exercise 3

Dina, Lila, and Amira pulled three consecutive numbers greater than 10 out of a magical hat. Dina said: "Wow! The digit sum of my number is equal to the sum of the digit sums of your numbers!" Which of the following statements must be true? Check all that apply.

(A) Dina's number is the sum of the numbers held by Lila and Amira.

(B) Dina's number is a repdigit.

(C) Lila's number has the same number of digits as Amira's number.

(D) Any three consecutive numbers have this property.

Exercise 4

Lila, Amira, and Dina each had 99 pennies they wanted to use for their artwork project. Their projects were different, so Lila gave Amira 19 of her pennies, Amira gave Dina 29 of her pennies, and Dina gave Lila 9 of her pennies. Which of the following graphs shows the number of pennies they each had after the exchanges?

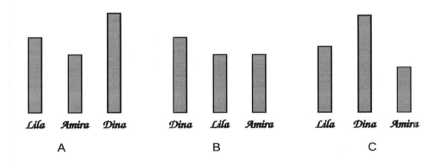

Exercise 5

The Djinn Jiskan will give Ali and Baba the key to the Sand Treasure if they can point out which of the following statements must be false:

(A) Four consecutive numbers can have a sum that is divisible by 4.

(B) Four consecutive numbers can have a sum that is divisible by 2.

(C) Four consecutive numbers can have a sum that is divisible by 3.

(D) Four consecutive numbers can have a sum that is divisible by 5.

Ali and Baba are treasure hunters, not mathematicians - please help them get the key!

Exercise 6

How many pairs of numbers ranging from 20 to 49 are multiples of one another?

Exercise 7

Two positive integers have a product that ends in three zeros and a sum of 261. One of the integers is an odd 3-digit integer, the other one is an even 3-digit integer, and neither of them ends in zero. What is the difference between the two integers?

Exercise 8

The Djinn wanted to meet Ali, who was in Cairo, and Baba, who was in Marrakech. The Djinn waited for them in the Seba oasis. The Djinn knew that Ali was always 45 minutes late for appointments while Baba always arrived 35 minutes early. If the Djinn wanted to meet them at exactly 8 AM, what time did he tell each of them to meet him at?

Exercise 9

Which number does K represent?

$$K + K + K + 333 = 2109$$

Exercise 10

Dina added 6 to a number and then multiplied the result by 11 to get 1111. What number did Dina start with?

Exercise 11

The digit product of a number ends in two zeros. The number has at least:

(**A**) 2 digits

(**B**) 3 digits

(**C**) 4 digits

(**D**) 5 digits

Exercise 12

A fifth of the students enrolled in Stephan's tennis classes is 4 students less than one third of the enrolled students. If each student has a weekly hour-long lesson, for how many hours does Stephan coach each week?

Exercise 13

The sum of a number and its digit sum is 117. How many numbers with this property are there?

(A) 0

(B) 1

(C) 2

(D) 116

Exercise 14

Jared had 1000 crickets and 7 boxes. He put crickets in each box without counting them. Jared asked several assistants to count the crickets and tell him what they found. After they did so, Jared began to suspect that several assistants did not know much math. Which of the following statements gave Jared this impression?

(A) "Each box contains an odd number of crickets."

(B) "Each box contains the same number of crickets."

(C) "There are 2 boxes with even numbers of crickets and 5 boxes with odd numbers of crickets."

(D) "The numbers of crickets in the boxes form a sequence of consecutive numbers."

Exercise 15

Which number should be removed from the list for the remaining numbers to have a sum that is divisible by 6?

$$61, \ 98, \ 81, \ 114, \ 83, 142$$

Exercise 16

What are the next two terms of the following sequence?

$$100, \ 101, \ 110, \ 111, \ 1000, \ 1001, \ 1010, \ 1011, \ 1100, \ \cdots$$

Exercise 17

What are the next two terms of the following sequence?

$$N, \ Y, \ YN, \ YY, \ YNN, \ YNY, \ YYN, \ YYY, \ YNNN, \ YNNY, \ \cdots$$

Exercise 18

Dina and Lila played a number game. One of them thought of a positive integer and the other one asked questions with yes or no answers to find out which number it was. Dina thought of a number and Lila asked the following questions:

1. Is it a 2 digit number?
2. Is it a 1 digit number?
3. Is it even?
4. Is it odd?
5. Is it a multiple of 5?

Lila then figured out which number it was.

1. Which number was it?
2. What were Dina's answers to the questions?
3. Which question could Lila have skipped?

Exercise 19

Write the next two terms in the following sequence:

$$1, \ 2, \ 6, \ 24, \ \cdots$$

Exercise 20

Alfonso received fresh pastry from Max, his baker. There were as many croissants as Danishes, and as many strudels as baklavas. There were four times as many croissants as strudels. The pastries came in boxes of 15 pieces each. Alfonso received 11 boxes, but one of them was not full. How many baklavas did he receive?

Exercise 21

Find every two digit number that decreases by 5 when its digits are reversed.

Exercise 22

In the following calendar, the dates hidden by the letters Q and S could have the same sum as the dates hidden by:

Sun	Mon	Tue	Wed	Thu	Fri	Sat
		P	W	Q		
		S	T	V		
		R	M	K		

(A) W and T

(B) V and R

(C) R and W

(D) M and P

Exercise 23

Alfonso has 14 boxes of onions lined up on a shelf of his store. The number of onions in each box differs from the number of onions in a neighboring box by either 1, 3, 7, or 13. The total number of onions in the 14 boxes cannot be:

(A) 47

(B) 128

(C) 113

(D) 59

Exercise 24

A 3-digit number is added to its digit sum and the result is 513. How many such numbers are there:

(A) 0

(B) 1

(C) 2

(D) 500

Exercise 25

The sum of 6 distinct positive integers is 19. The largest number is:

(A) 6

(B) 8

(C) 9

(D) 10

(E) there is no solution

Exercise 26

The product of two numbers is 203. If we subtract a number from one of the two numbers, their product becomes 175. Which number was subtracted?

Exercise 27

Two of four consecutive numbers have a sum of 21. The sum of the other two cannot be:

(A) 17

(B) 21

(C) 23

(D) 25

Exercise 28

Lila had a paper rectangle with length 16 units and width 12 units. She repeatedly folded it in half until she could no longer obtain integer length sides. How many folds did she make?

SOLUTIONS TO PRACTICE ONE

Solution 1

1. The largest integer with all different even digits is 86420.

2. The largest integer with all different odd digits is 97531.

3. The largest integer with all different digits is 9876543210.

4. The next integer with all different digits after 69785 is 69801.

5. The largest 6-digit integer with a digit product of 12 is 621111.

6. The largest palindrome written with 3 5s and 6 3s is 533353335.

7. The largest 6 digit integer with a digit sum of 12 is 930000.

8. The smallest 6 digit integer with a digit sum of 12 is 100029.

Exercise 2

All the digits of a number A are different. A number can be formed using different digits that are each larger than the digits of A by one. What is the largest such number A?

Solution 2

$A = 876543210$. The other number is formed with the digits 1, 2, 3, 4, 5, 6, 7, 8, and 9.

Exercise 3

All the digits of a number B are different. A number can be formed using different digits that are the double of the digits of B. What is the largest such number B?

Solution 3

The number B can only have digits whose doubles are also digits: 0, 1, 2, 3, and 4. The largest such number has only 5 digits: 43210.

Exercise 4

Stephan, the tennis coach, wrote an invoice for one of his clients. He noticed that the number of the invoice was 91087 and asked himself: "After how many more invoices will there be another invoice number in which all the digits are different?"

Solution 4

The next number with different digits is 91203. Stephan will have to write another $91203 - 91087 = 116$ invoices.

Exercise 5

1. How many digits are multiples of 3?
2. How many digits are multiples of 4?
3. How many digits are multiples of 5?
4. How many digits are multiples of 7?

Solution 5

1. There are 4 digits that are multiples of 3: 0, 3, 6, and 9.
2. There are 3 digits that are multiples of 4: 0, 4, and 8.
3. There are 2 digits that are multiples of 5: 0 and 5.
4. There are 2 digits that are multiples of 7: 0 and 7.

Exercise 6

The digits m, n, and s can be sorted in increasing order like this: $m > n > s$. With these digits, we can form 6 three-digit numbers. Which of the following cryptarithms are true? Check all that apply.

(A) $msn > nms > smn$

(B) $nms > mns > smn$

(C) $mns > nsm > snm$

(D) $nms > nsm > msn$

Solution 6

Work on a concrete example, such as:

$$321 > 312 > 231 > 213 > 132 > 123$$

Replace 3 with m, 2 with n, and 1 with s. The 6 numbers can be ordered as follows:

$$mns > msn > nms > nsm > smn > snm$$

By comparing each answer choice with these inequalities, we conclude that (A) and (C) are true, while (B) and (D) are false.

Exercise 7

The number 44444 can be written as the sum of two integers that differ by 6666. Find the value of the smaller integer.

Solution 7

$$44444 \div 2 = 22222$$

$$22222 + 3333 = 25555$$

$$22222 - 3333 = 18889$$

The smaller integer is 18889.

Exercise 8

If the tens digit of a number is 4 times the units digit and the hundreds digit is twice the tens digit, is the number even or odd?

Solution 8

8 times the units digit must also be a digit. The only digit that satisfies this condition is 1. The number ends with the digits 841. The number is odd.

Exercise 9

Some numbers have a digit product of 30 and, when multiplied by 5, produce a number that ends in zero. What is the largest 6-digit number of this kind?

Solution 9

Factor 30 to see which digits other than zero the number can have:

$$30 = 2 \times 3 \times 5$$

The number has digits 2, 3, and 5 and 3 digits of 1, or digits 6, 5 and 4 digits of 1.

If 5 times the number ends in zero, then the number must end with an even digit. Two types of numbers are possible:

- 511116 and all the possible permutations of its first 5 digits
- 531112 and all the possible permutations of its first 5 digits

Of all these numbers, 531112 is the largest.

Exercise 10

What are the first three digits of the product: 999999999999×6?

Solution 10

Write 999999999999 as $1000000000000 - 1$ and use the distributive property:

$$(1000000000000 - 1) \times 6 = 6000000000000 - 6 = 5999999999994$$

The first three digits are 599.

Exercise 11

Three consecutive numbers have digit sums of 18, 1, and 2. What is the digit sum of their sum?

Solution 11

The only consecutive numbers with these digit sums are: 99, 100, and 101. Their sum is $99 + 100 + 101 = 300$ and the digit sum of their sum is 3.

Exercise 12

How many positive integers with non-zero digit products have digit sums of 1, 2, or 3?

Solution 12

Numbers that have a digit sum of 1 and a non-zero digit product: 1.

Numbers that have a digit sum of 2 and a non-zero digit product: 2, 11.

Numbers that have a digit sum of 3 and a non-zero digit product: 3, 12, 21, and 111.

In total, there are 7 numbers that satisfy the conditions.

Exercise 13

What is the smallest positive integer with a digit sum of 20?

Solution 13

The smallest number with a digit sum of 20 can be obtained by using the largest digits possible. If the digits are large, then there are less of them. Fewer digits mean smaller place values. Therefore, use as many 9s as possible and place them behind the remainder, which is smaller than 9 and should be given the largest place value.

Since $20 = 9+9+2$, the smallest number with a digit sum of 20 is 299.

Exercise 14

How many digits does the smallest positive integer with a digit sum of 203 have?

Solution 14

Since $203 = 9 \times 22 + 5$, the smallest number with a digit sum of 203 is:

$$5 \underbrace{999 \cdots 99}_{22 \text{ digits}}$$

The number has 23 digits.

Exercise 15

How many occurences of the digit 9 are there in the smallest positive integer with a digit sum of 1003?

Solution 15

Since:

$$1003 = 9 \times 111 + 4$$

the smallest positive integer with a digit sum of 1003 has 111 digits of 9:

$$4 \underbrace{999 \cdots 99}_{111 \text{ digits}}$$

The correct answer is (C).

Exercise 16

What is the remainder when we divide each of the following by 3?

$$10, \ 100, \ 1000, \ 10000, \ 100000$$

Solution 16

All the remainders are equal to 1.

Exercise 17

What is the remainder when we divide each of the following by 9?

10, 100, 1000, 10000, 100000

Solution 17

All the remainders are equal to 1.

Exercise 18

What is the remainder when we divide each of the following by 11?

10, 100, 1000, 10000, 100000

Solution 18

The remainders form the sequence: 10, 1, 10, 1, 10.

SOLUTIONS TO PRACTICE TWO

Solution 1

Write three more terms to the following sequences:

1. 10, 11, 100, 101, 110, 111, 1000, **1001, 1010, 1011**, $\cdots$
2. 3, 4, 3, 4, 5, 6, 5, 6, 7, **8, 7, 8**, $\cdots$
3. 1, 2, 4, 7, 11, **16, 22, 29**, $\cdots$
4. 11, 22, 66, 264, **1320, 7920, 55440**, $\cdots$

The sequences are formed using the following rules:

1. The rightmost digit changes every term. The tens digit changes every 2 terms. The hundreds digit changes every 4 terms. The thousands digit changes every 8 terms, etc.
2. Each pair of consecutive (odd, even) numbers, starting with (3, 4), is repeated twice.
3. Add each term to its rank to obtain the next term.
4. Multiply each term by the rank of the next term to obtain the next term.

Exercise 2

Lila divided the numbers in the following list by 10 and placed the remainders in boxes, one number per box. How many boxes did she use?

71, 99, 32, 401, 63, 27, 49, 111, 60, 94, 85, 868, 18, 19

Solution 2

When we divide by 10, the remainder of a number is its last digit. There are as many boxes as different last digits. The only digit that is not used as a units digit in the set of numbers is the digit 6. All other digits are used as last digits. Dina will have to use 9 boxes.

58

Exercise 3

Dina had to divide positive consecutive integers starting from 1 by 7 and write down the sequence formed by the remainders. Soon, Dina realized she no longer needed to write more of the sequence: she was able to predict any future term. For example, Dina could predict the 45th term in the sequence. Which one of the following numbers matches Dina's prediction?

Solution 3

The sequence of remainders is:

$$1, \ 2, \ 3, \ 4, \ 5, \ 6, \ 0, \ 1, \ 2, \ \cdots$$

The sequence is formed by groups of 7 remainders (1, 2, 3, 4, 5, 6, 0) which repeat indefinitely. To figure out the 45th term of this sequence, divide 45 by 7:

$$45 = 6 \times 7 + 3$$

This means that there are 6 complete groups of 7 remainders and 3 remainders left over: 1, 2, 3. The 45th term is equal to 3. The correct answer is (B).

Exercise 4

What is the remainder when we divide the following number, in which some digits have been hidden by symbols, by 100?

$$7\spadesuit\diamondsuit8\clubsuit3\diamondsuit\heartsuit66\diamondsuit2\clubsuit\clubsuit\heartsuit532$$

Solution 4

When we divide by 100, the remainder is the number formed by the last two digits of the dividend. For this number, the remainder is 32. The values of the hidden digits do not play a role.

Exercise 5

Amira and Dina used some chalk to create a new game in the driveway. They drew 6 squares in a line and marked them with the letters A, B, C, D, M, and N. Dina started at A and jumped to B, then to C, and so on until square N. On square N, she turned around and jumped from square to square in the opposite direction. The same thing happened when she reached square A again, and so on. Which letter was on the square Dina landed on after her 28th jump?

Solution 5

Dina will jump onto square C.

All the jumps that are even multiples of 5 land on square A and all the jumps that are odd multiples of 5 land on square N. Therefore, Dina lands on N after the 25th jump and on C after the 28th jump.

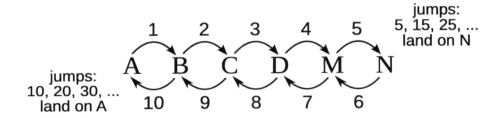

Exercise 6

Three bunnies, Keef, Leef, and Meef, stand in line with Keef first and Meef last. They start jumping, repeating the following rules:

- the last jumps over the middle one,
- the middle one jumps over the first one.

After how many jumps will Meef be last in line again?

Solution 6

Meef will be the last in line again after 5 jumps.

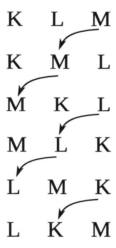

Exercise 7

Dina says "42" and Lila replies "6." Dina says "901" and Lila replies "10". Dina says "3" and Lila replies "3." Dina says "181" and Amira must guess Lila's reply.

Solution 7

Amira noticed that Lila's number is the digit sum of Dina's number. Amira thinks that Lila will reply "10" since $S(181) = 1 + 8 + 1 = 10$.

Exercise 8

A machine takes a number as an input and outputs a number related to it. This is a list of inputs and outputs:

input	151	78	100	91
output	5	56	0	9

What will the machine output if the input is 345?

Solution 8

The machine outputs the digit product of the input. If the input is 345 the output will be $P(345) = 3 \times 4 \times 5 = 60$.

Exercise 9

What is the next term in the sequence?

$$1, \ 2, \ 2, \ 4, \ 8, \ 32, \ \cdots$$

Solution 9

Except for the first two terms, each term of the sequence is the product of the two previous terms. The next term is $8 \times 32 = 256$.

Exercise 10

If Dina guesses the next term of each of the following sequences correctly, she gets as many points as the term she guessed. She starts the game with zero points. If Dina scored 18 points, how many times did she guess correctly?

(A) 1

(B) 2

(C) 3

(D) 4

Solution 10

The next terms have been bolded:

1. Z, 1, Y, 2, X, 3, W, **4**, $\cdots$

2. 1, 1, 2, 3, 5, 8, **13**, $\cdots$

3. 1, 7, 5, 11, 9, 15, **13**, $\cdots$

4. 7, 9, 3, 1, 7, 9, 3, **1**, $\cdots$

If Dina scored 18 points she must have 3 correct answers. She got either question 2 or question 3 wrong.

Exercise 11

A sequence is generated by the following machines.

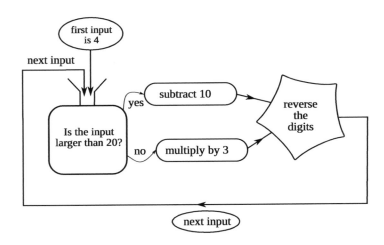

Calculate the first 10 inputs.

Solution 11

1. The first input is 4 which is smaller than 20.
   Multiply 4 by 3 and get 12.
   Reverse the digits of 12 and get 21.

2. The second input is 21 which is larger than 20.
   Subtract 10 and get 11.
   Reverse the digits and get 11.

3. The 3rd input is 11 which is smaller than 20.
   Multiply by 3 and get 33.
   Reverse the digits and get 33.

4. The 4th input is 33 which is larger than 20.
   Subtract 10 and get 23.
   Reverse the digits and get 32.

5. The 5th input is 32 which is larger than 20.
   Subtract 10 and get 22.
   Reverse the digits and get 22.

6. The 6th input is 22 which is larger than 20.
   Subtract 10 and get 12.
   Reverse the digits and get 21.

7. The 7th input is 21 which is larger than 20.
   Subtract 10 and get 11.
   Reverse the digits and get 11.

8. The 8th input is 11 which is smaller than 20.
   Multiply by 3 and get 33.
   Reverse the digits and get 33.

9. The 9th input is 33 which is larger than 20.
   Subtract 10 and get 23.
   Reverse the digits and get 32.

10. The 10th input is 32.

The first 10 terms of the sequence generated by the machines are:

$$4, \ 21, \ 11, \ 33, \ 32, \ 22, \ 21, \ 11, \ 33, \ 32$$

Exercise 12

Dina cut a square into four small squares. Then, she cut one of the small squares into four squares. She repeated this operation 10 times, each time cutting into four one of the smallest squares available at the time. How many squares did she end up with?

Solution 12

At each operation, four squares of smaller size are produced. Three of them are left untouched while the fourth is cut into four. Therefore, 3 same size squares are produced by each operation, except for the last operation which produces 4 same size squares. An example with 3 operations is shown in the figure:

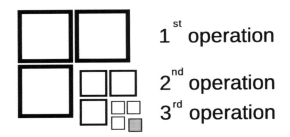

After 10 operations, the number of squares Dina ends up with is:

$$3 \times 9 + 4 = 31$$

Solutions to Practice Three

Solution 1

1. $1 + 2 + 3 + 4 + 5 + 6 + 7 = \dfrac{7 \times 8}{2} = 4 \times 7 = \mathbf{28}$

2. $1 + 2 + 3 + 4 + \cdots + 10 = \dfrac{10 \times 11}{2} = 5 \times 11 = \mathbf{55}$

3. $1 + 2 + 3 + 4 + \cdots + 50 = \dfrac{50 \times 51}{2} = 25 \times 51 = \mathbf{1275}$

4. $1 + 2 + 3 + 4 + \cdots + 97 = \dfrac{97 \times 98}{2} = 97 \times 49 = 97 \times 50 - 97 = 4850 - 97 = \mathbf{4753}$

Solution 2

1. $3 + 4 + 5 + 6 = 1 + 2 + 3 + 4 + 5 + 6 - 1 - 2 = 6 \times 72 - 3 = 21 - 3 = \mathbf{18}$

2. $5 + 6 + 7 + 8 + 9 + 10 + 11 + 12 + 13 + 14 + 15 = \dfrac{15 \times 16}{2} - \dfrac{4 \times 5}{2} = 15 \times 8 - 5 \times 2 = 120 - 10 = \mathbf{110}$

3. $3 + 4 + 5 + \cdots + 99 = \dfrac{99 \times 100}{2} - 1 - 2 = 99 \times 50 - 3 = 4950 - 3 = \mathbf{4947}$

4. $10 + 11 + 12 + \cdots + 91 = \dfrac{91 \times 92}{2} - \dfrac{9 \times 10}{2} = 91 \times 46 - 9 \times 5 = 4186 - 45 = \mathbf{4141}$

5. $101 + 102 + 103 + \cdots + 150 = \dfrac{150 \times 151}{2} - \dfrac{100 \times 101}{2} = 75 \times 151 - 50 \times 101 = 11325 - 5050 = \mathbf{6275}$

Solution 3

1. $2+4+6+8+10 = 2 \times (1+2+3+4+5) = 2 \times \dfrac{5 \times 6}{2} = 5 \times 6 = \mathbf{30}$

2. $2+4+6+\cdots+20 = 2 \times (1+2+3+\cdots+10) = 2 \times \dfrac{10 \times 11}{2} = \mathbf{110}$

3. $20 + 40 + 60 + \cdots + 200 = 20 \times (1 + 2 + 3 + \cdots 10) = \mathbf{1100}$

4. $10+12+14+\cdots+20 = 2 \times (5+6+\cdots+10) = 2 \times (55-10) = \mathbf{90}$

5. $100+102+104+\cdots+110 = 100+100+2+100+4+\cdots+100+10 = 6 \times 100 + 2 + 4 + \cdots + 10 = 600 + 30 = \mathbf{630}$

Solution 4

1. $10+20+30+\cdots+100 = 10 \times (1+2+3+\cdots+10) = 10 \times 55 = \mathbf{550}$

2. $50 + 60 + 70 + \cdots + 100 = 10 \times (5+6+\cdots+10) = 10 \times 45 = \mathbf{450}$

3. $1 + 3 + 5 + 7 + 9 + 11 + 13 = 7 \times 7 = \mathbf{49}$

4. $101 + 103 + 105 + \cdots + 111 = 100 \times 6 + (1+3+5+7+9+11) = 600 + 36 = \mathbf{636}$

5. $3+6+9+12+\cdots+36 = 3 \times (1+2+3+\cdots+12) = 3 \times \dfrac{12 \times 13}{2} = 3 \times 6 \times 13 = 3 \times 78 = \mathbf{234}$

Exercise 5

A list of consecutive integers has a sum of 4. How many integers are there in this list?

Solution 5

There are eight terms in the list.

4 is not a triangular number. Therefore, the list consists of positive and negative numbers that cancel each other out plus the number 4:

$$-3, \ -2, \ -1, \ 0, \ 1, \ 2, \ 3, \ 4$$

Exercise 6

A list of consecutive integers has a sum of 11. How many integers are there in this list? Is there more than one answer?

Solution 6

There are two solutions 2 and 22.

The list could consist of two numbers: 5 and 6.

Also, the list could consist of positive and negative numbers that cancel each other plus the number 11. In this case, there are 22 numbers in the list:

$$-10, \ -9, \ -8, \ \cdots, \ 0, \ 1, \ 2, \cdots, \ 9, \ 10, \ 11$$

Exercise 7

Mira is a dog breeder. Each of her Afghan hounds was born in a different consecutive year. When the sum of the ages of the hounds was 9, how old was the oldest dog? Is there more than one solution?

Solution 7

Since $9 = 3 \times 3$, we have a solution with an odd number of dogs:

$$9 = 2 + 3 + 4$$

There is also one solution with an even number of dogs:

$$9 = 4 + 5$$

There are 2 solutions in total.

Exercise 8

The sum of three consecutive numbers is 31038. What is the middle number?

Solution 8

Strategy 1:

The smallest number is smaller than the middle number by 1. The largest number is larger than the middle number by 1. If we transfer a 1 from the largest number to the smallest number, the three numbers will be equal. It follows that dividing the sum by 3 will give us the middle number: $31038 \div 3 = 10346$.

Strategy 2:

Imagine the numbers listed in increasing order. The second number is greater than the first number by 1 and the third number is greater than the first number by 2. Subtract 3 from the sum of the numbers. The result will be three times the smallest number: $31038 - 3 = 31035$. Divide 31035 by 3 to find the smallest number: $31035 \div 3 = 10345$. The middle number is 10346.

Exercise 9

Write the number 18105 as a sum of 6 consecutive multiples of 5.

Solution 9

Six consecutive multiples of 5 could be written like this:

$$n - 10$$
$$n - 5$$
$$n$$
$$n + 5$$
$$n + 10$$
$$n + 15$$

When adding these multiples together, we notice that, by subtracting 15 from the sum, we get n added to itself 6 times. We obtain the value of n in the following manner:

$$18105 - 15 = 18090$$
$$18090 \div 6 = 3015$$

The numbers are: 3005, 3010, 3015, 3020, 3025, and 3030.

Exercise 10

A fisherman's boat capsized in a storm and its catch of fish spilled onto a remote beach. A penguin came by and started eating fish at a rate of one fish every 2 minutes. As he was starting the second fish, another penguin arrived and started eating fish at the same rate. A new penguin arrived and began to eat fish each time the previous penguin started its second fish. How long did it take the penguins to eat the first 21 fish?

Solution 10

The first penguin ate 1 fish.

After 2 minutes, two penguins ate 2 fish.

After another 2 minutes, three penguins ate 3 fish.

71

This goes on until 21 fish are eaten. 21 is a triangular number:

$$1 + 2 + 3 + 4 + 5 + 6 = 21$$

21 fish were eaten in total, after the sixth penguin joined in. This happened $2 \times 6 = 12$ minutes after the first penguin started eating fish.

Exercise 11

How many digits does the following number have? Assume the pattern is the same to the end.

$$1223334444 \cdots 99$$

Solution 11

For each digit, there are as many digits of that kind as the value of the digit. There is 1 digit of 1, 2 digits of 2, etc. Therefore, the total number of digits is:

$$1 + 2 + 3 + \cdots + 9 = 9 \times 10 \div 2 = 45$$

Exercise 12

Dina was completing a worksheet and had to compute the sum of 87 consecutive numbers. Arbax, the Dalmatian, played with the sheet and managed to damage it so that only the middle term of the sum was still legible: 100. What was the largest term of the sum?

Solution 12

If 100 was the middle number, there were 86 other numbers in the sum, 43 of them smaller than 100 and 43 of them larger than 100. The largest term in the sum was $100 + 43 = 143$.

Exercise 13

How many pairs of digits (a, b), with $a > b$, are there that satisfy the following cryptarithm? (The same letter represents the same digit. Different letters represent different digits.)

$$aaa + bbb = ccc$$

Solution 13

Only digits that do not produce a carryover when added can satisfy the cryptarithm.

If a is 9, there is no solution.
If a is 8, b can be 1.
If a is 7, b can be 1 or 2.
If a is 6, b can be 1, 2, or 3.
If a is 5, b can be 1, 2, 3, or 4.
If a is 4, b can be 1, 2, or 3. (Remember that a is larger than b!)
If a is 3, b can be 1 or 2.
If a is 2, b can be 1.
If a is 1, there is no solution.

The total number of pairs is equal to:

$$1 + 2 + 3 + 4 + 3 + 2 + 1 = 16$$

There are 16 pairs of digits in total.

Solution 14

Compute the sums:

1.

$$
\begin{aligned}
3 + 6 + \cdots + 99 &= 3 \times (1 + 2 + \cdots + 33) \\
&= 3 \times \frac{33 \times 34}{2} \\
&= 3 \times 33 \times 17 \\
&= 99 \times 17 \\
&= 100 \times 17 - 17 \\
&= 1700 - 17 \\
&= \mathbf{1683}
\end{aligned}
$$

2.

$$
\begin{aligned}
4 + 8 + 12 + \cdots + 100 &= 4 \times (1 + 2 + \cdots + 25) \\
&= 4 \times \frac{25 \times 26}{2} \\
&= 50 \times 26 \\
&= \mathbf{1300}
\end{aligned}
$$

3.

$$
\begin{aligned}
5 + 10 + 15 + \cdots + 100 &= 5 \times (1 + 2 + \cdots + 20) \\
&= 5 \times \frac{20 \times 21}{2} \\
&= 50 \times 21 \\
&= \mathbf{1050}
\end{aligned}
$$

4.

$$
\begin{aligned}
6 + 12 + 18 + \cdots + 180 &= 6 \times (1 + 2 + \cdots + 30) \\
&= 6 \times \frac{30 \times 31}{2} \\
&= 90 \times 31 \\
&= \mathbf{2790}
\end{aligned}
$$

Exercise 15

What is the value of k?

$$
1 + 2 + 3 + \cdots + k = 820
$$

Solution 15

$k(k+1)$ equals 2×820. Factor this number and group factors so as to obtain a product of two consecutive integers:

$$
\begin{aligned}
820 \times 2 &= 2 \times 2 \times 2 \times 5 \times 41 \\
820 \times 2 &= 40 \times 41
\end{aligned}
$$

$k = 40$.

Solutions to Practice Four

Solution 1

Examine the number formed by the last two digits:

- 38 is not a multiple of 4, therefore 138 is not divisible by 4;
- 42 is not a multiple of 4, therefore 142 is not divisible by 4;
- 18 is not a multiple of 4, therefore 118 is not divisible by 4;
- 42 is not a multiple of 4, therefore 442 is not divisible by 4;
- 98 is not a multiple of 4, therefore 98 is not divisible by 4;
- 46 is not a multiple of 4, therefore 146 is not divisible by 4;
- 86 is not a multiple of 4, therefore 886 is not divisible by 4;
- 14 is not a multiple of 4, therefore 2014 is not divisible by 4;
- 76 is a multiple of 4, therefore 76 is divisible by 4.

Only one number in the list is divisible by 4.

Solution 2

Use the sum of the digits to decide whether the number is divisible by 3:

- $S(96) = 9 + 6 = 15$ and $S(15) = 1 + 5 = 6$, therefore 96 is divisible by 3;
- $S(144) = 1 + 4 + 4 = 9$, therefore 144 is divisible by 3;
- $S(32) = 3 + 2 = 5$, therefore 32 is not divisible by 3;
- $S(87) = 8 + 7 = 15$ and $S(15) = 1 + 5 = 6$, therefore 87 is divisible by 3;
- 0 is divisible by any number, including 3;
- $S(100) = 1$, therefore 100 is not divisible by 3;
- $S(51) = 5 + 1 = 6$, therefore 51 is divisible by 3;
- $S(37) = 3 + 7 = 10$ and $S(10) = 1 + 0$, therefore 37 is not divisible

by 3;

- $S(57) = 5 + 7 = 12$ and $S(12) = 1 + 2 = 3$, therefore 57 is divisible by 3;

- $S(11) = 1 + 1 = 2$, therefore 11 is not divisible by 3;

6 of the numbers in the list are divisible by 3.

Solution 3

Apply the digit sum criterion to find out if each number is divisible by 3 but not divisible by 9:

- $S(15) = 1 + 5 = 6$. 15 is divisible by 3 but not by 9.
- $S(45) = 4 + 5 = 9$. 45 is divisible both by 3 and by 9.
- $S(32) = 3 + 2 = 5$. 32 is not divisible by 3 or by 9.
- $S(93) = 9 + 3 = 12$ and $S(12) = 1 + 2 = 3$. 93 is divisible by 3 but not by 9.
- $S(111) = 1 + 1 + 1 = 3$. 111 is divisible by 3 but not by 9.
- $S(38) = 3 + 8 = 11$ and $S(11) = 1 + 1 = 2$. 38 is not divisible by 3 or by 9.
- 0 is a multiple of any number: it is divisible by both 3 and 9.
- $S(42) = 4 + 2 = 6$. 42 is divisible by 3 but not by 9.
- $S(108) = 1 + 8 = 9$. 108 is divisible both by 3 and by 9.
- $S(72) = 7 + 2 = 9$. 72 is divisible by both 3 and by 9.
- $S(69) = 6 + 9 = 15$ and $S(15) = 1 + 5 = 6$. 69 is divisible by 3 but not by 9.
- $S(174) = 1 + 7 + 4 = 12$ and $S(12) = 1 + 2 = 3$. 69 is divisible by 3 but not by 9.
- $S(153) = 1 + 5 + 3 = 9$. 153 is divisible both by 3 and by 9.

Exercise 4

Find the largest number of the form 2♠8 that is divisible by 4 but not by 3.

Solution 4

A number is divisible by 4 if its last two digits form a number divisible by 4. The choices for the last two digits are, therefore: 08, 28, 48, 68, and 88. Now we calculate the digit sum of each choice to find the choices that are not divisible by 3:

$$
\begin{aligned}
S(208) &= 10, & S(10) = 1 + 0 = 1 \\
S(228) &= 12, & S(12) = 1 + 2 = 3 \\
S(248) &= 14, & S(14) = 1 + 4 = 5 \\
S(268) &= 16, & S(16) = 1 + 6 = 7 \\
S(288) &= 18, & S(18) = 1 + 8 = 9
\end{aligned}
$$

The numbers that satisfy the conditions are: 208, 248 and 268. The largest one is 268.

Solution 5

Use the alternating digit sum to find out which numbers are divisible by 11:

- $S(141) = 1 - 4 + 1 = -2$. 141 is not divisible by 11.
- $S(101) = 1 - 0 + 1 = 2$. 101 is not divisible by 11.
- $S(1001) = 1 - 0 + 0 - 1 = 0$. 1001 is divisible by 11.
- $S(154) = 4 - 5 + 1 = 0$. 154 is divisible by 11.
- $S(1067) = 7 - 6 + 0 - 1 = 0$. 1067 is divisible by 11.
- $S(111) = 1 - 1 + 1 = 1$. 111 is not divisible by 11.
- $S(187) = 7 - 8 + 1 = 0$. 187 is divisible by 11.
- $S(209) = 9 - 0 + 2 = 11$. 209 is divisible by 11.
- $S(134) = 4 - 3 + 1 = 2$. 134 is not divisible by 11.

Exercise 6

Which of the following repunits are divisible by 11?

$$11, \; 111, \; 1111, \; 11111, \; 111111, \; 1111111$$

Fill in the blanks:

Repunits with .... number of digits are divisible by 11.

Does the rule you just found hold for any repdigit or only for repunits?

Solution 6

11, 1111, and 111111 are divisible by 11.

111, 11111, and 1111111 are not divisible by 11.

Repunits with an even number of digits are divisible by 11.

Repunits with an odd number of digits are not divisible by 11.

Since repdigits are multiples of repunits, it follows that, if a repunit is divisible by some number, so are all the repdigits with the same number of digits.

Solution 7

1. $1\underbrace{11\cdots1}_{111 \text{ digits}}1$ is a repunit with 113 digits, and is not a multiple of 11.

2. $4\underbrace{44\cdots4}_{444 \text{ digits}}4$ is a repdigit with 446 digits, and is a multiple of 11.

3. $7\underbrace{77\cdots7}_{777 \text{ digits}}7$ is a repdigit with 779 digits, and is not a multiple of 11.

4. $6\underbrace{66\cdots6}_{666 \text{ digits}}6$ is a repdigit with 668 digits, and is a multiple of 11.

Exercise 8

Dina, Lila, and Amira played a game in which they received a number with hidden digits. They had to find the hidden digits so that the number was divisible by both 2 and 3. Each of them was then allowed to add the numbers they found to their score. If the number they started with was 25♣, which of the following statements is true?

(A) Dina won because she got 252, the highest possible score.

(B) Lila won because she got 258, the highest possible score.

(C) Amira won because she got 510, the highest possible score.

Solution 8

If the number must be divisible by 2, then the ♣ can be either 0, 2, 4, 6, or 8. For the number to also be a multiple of 3, its digit sum must be a multiple of 3. The digit sum is $S(25♣) = 2 + 5 + ♣$ and can have the values 7, 9, 11, 13, or 15. Of these, only two are multiples of 3. Therefore, there are two numbers that satisfy: 252 and 258. While 252 and 258 are possible scores, they are not the highest possible. The highest possible score, 510, was obtained by Amira, who found both solutions. The correct answer choice is (C).

Exercise 9

Alfonso, the grocer, has a list of all the products he received from a delivery van and wants to match this list to the order he placed. Unfortunately, the driver of the van spilled some coffee on the list and some of the digits are no longer legible:

Potatoes	3 ● 4	lbs	
Lemons	1 6 0	lbs	
Beets	● 5 9	lbs	

Potatoes come in bags of 3 lbs each, lemons come in bags of 2 lbs each, and beets come in bags of 7 lbs each. Moreover, Alfonso is sure he ordered a smaller quantity of beets than of potatoes. Which of the following cannot be the total weight of the merchandise Alfonso received?

(A) 743 lbs

(B) 773 lbs

(C) 803 lbs

(D) 809 lbs

Solution 9

The weight of the potatoes must be a multiple of 3. The digit sum of the number is $S(3\Diamond 4) = 3 + \Diamond + 4 = 7 + \Diamond$. The illegible digit could be $2, 5$ or 8. Therefore, the weight of the potatoes could be either: 324 lbs, 354 lbs, or 384 lbs.

The weight of the beets must be a multiple of 7. We do not have a rule of divisibility by 7 that is simple enough, so we will simply divide by 7. To do this more easily, first write the number in expanded form:

$$\Diamond 59 = \Diamond 00 + 59 = \Diamond \times 100 + 59$$

59 gives a remainder 3 when divided by 7. Therefore, $\Diamond \times 100$ must give a remainder of 4 when divided by 7. Since 100 gives a remainder of 2 when divided by 7, it follows that 200 will give a remainder of 4. The number we are looking for could be 259. Is it the only possible solution?

If each 100 gives a remainder of 2, then:

- 200 gives a remainder of 4.
- 300 gives a remainder of 6.
- 400 gives a remainder of 1.

Already we are above the weight of the potatoes and we do not need to look any further!

Remembering that there are also 160 lbs of lemons, the total weight of the shipment can be: $324 + 259 + 160 = 743$ lbs, $354 + 259 + 160 = 773$

lbs, or $384 + 259 + 160 = 803$ lbs. Answer choice (D) is the correct answer.

Exercise 10

Arbax, the Dalmatian, must guess the hidden digits for each number so that it is divisible by 3. For each number, Arbax may give one possible answer. For each correct answer, Arbax gets as many bones as the number he guessed. What is the largest number of bones Arbax can win?

1. 6♢4
2. ♣77
3. 11♠
4. 8♡1
5. 22♢

Solution 10

1. 6♢4 has solutions: 624, 654, and 684
2. ♣77 has solutions: 177, 477, and 777
3. 11♠ has solutions: 111, 114, and 117
4. 8♡1 has solutions: 801, 831, 861, and 891
5. 22♢ has solutions: 222, 225, and 228

The largest number of bones Arbax can get is:

$$684 + 777 + 117 + 891 + 228 = 2697$$

That's a lot of bones!

Exercise 11

Dina had to find all the numbers of the form $a55b$, where a and b are unknown digits, that are divisible by 15. How many such numbers are there?

Solution 11

The number must he divisible by both 3 and 5.

The number is divisible by 5 if b is 0 or 5.

Case 1: If $b = 0$, the digit sum is $S(a550) = 10 + a$ and a must be chosen so as to make the digit sum a multiple of 3. There are three possible values for a: 2, 5, and 8. So far, Dina has found three numbers.

Case 2: If $b = 5$, the digit sum is $S(a555) = 15 + a$ and a must be chosen so as to make the digit sum a multiple of 3. There are four possible values for a: 0, 3, 6, and 9. Dina has found another four numbers.

In total, there are seven numbers that satisfy the requirements.

Exercise 12

Alfonso received a shipment of onions. He packaged an equal number of onions in each bag and had some onions left over at the end. The sum of the number of onions in one bag and the number of bags was 21. The sum of the number of bags and the number of onions left over was 19. The sum of the number of onions in one bag and the number of onions left over was 10. How many onions were there in the shipment?

Solution 12

Denote the number of bags by B.
Denote the number of onions per bag by m.
Denote the number of onions left over by l.
Denote the total number of onions in the shipment by N.
We have to reconstitute the integer division:

$$N = B \times m + l$$

We can find B, m, and l from the information about their sums:

$$
\begin{aligned}
B + m &= 21 \\
B + l &= 19 \\
l + m &= 10
\end{aligned}
$$

We notice that, if we add all the equations, B, m, and l will each appear twice:

$$2 \times B + 2 \times m + 2 \times l = 21 + 19 + 10 = 50$$

Therefore,

$$B + m + l = 25$$

Since $B + m = 21$, l must be equal to 4.
Since $B + l = 19$, m must be equal to 6.
Since $l + m = 10$, B must be equal to 15.

The total number of onions in the shipment was:

$$15 \times 6 + 4 = 94$$

SOLUTIONS TO PRACTICE FIVE

Solution 1

105	3
35	5
7	7
1	

105 = 3 x 5 x 7

42	2
21	3
7	7
1	

42 = 2 x 3 x 7

1331	11
121	11
11	11
1	

1331 = 11 x 11 x 11

168	2
84	2
42	2
21	3
7	7
1	

168 = 2 x 2 x 2 x 3 x 7

352	2
176	2
88	2
44	2
22	2
11	11
1	

352 = 2 x 2 x 2 x 2 x 2 x 11

504	2
252	2
126	2
63	3
21	3
7	7
1	

504 = 2 x 2 x 2 x 3 x 3 x 7

Solution 2

First, estimate the square roots:

1. 151 is between $121 = 11 \times 11$ and $169 = 13 \times 13$. The largest prime to use is 11.

2. 127 is between $121 = 11 \times 11$ and $169 = 13 \times 13$. The largest prime to use is 11.

3. 191 is between $169 = 13 \times 13$ and $196 = 14 \times 14$. The largest prime to use is 13.

4. 257 is between $225 = 15 \times 15$ and $256 = 16 \times 16$. The largest prime to use is 13.

5. 233 is between $225 = 15 \times 15$ and $256 = 16 \times 16$. The largest prime to use is 13.

6. 331 is between $324 = 18 \times 18$ and $361 = 19 \times 19$. The largest prime to use is 17.

7. 499 is between $484 = 22 \times 22$ and $529 = 23 \times 23$. The largest prime to use is 19.

For each number, use simple divisibility rules to determine whether 2, 3, 5 or 11 are prime divisors of the number. Continue as follows:

1. For 151 attempt division by: 7.

2. For 127 attempt division by: 7.

3. For 191 attempt divisions by: 7 and 13.

4. For 257 attempt divisions by: 7 and 13.

5. For 233 attempt divisions by: 7 and 13.

6. For 331 attempt divisions by: 7, 13, and 17.

7. For 499 attempt divisions by: 7, 13, 17, and 19.

Solution 3

    1. 493 is prime.

    2. 499 is prime.

    3. 593 is prime.

    4. $323 = 17 \times 19$

    5. $1007 = 19 \times 53$

    6. $731 = 17 \times 43$

    7. $481 = 13 \times 37$

Exercise 4

How many 3-digit prime numbers can be formed using each of the digits 1, 3, and 5 once?

Solution 4

None. Compute the digit sum:

$$S(1,3,5) = 1 + 3 + 5 = 9$$

to find out that the number would be divisible by 3 (and by 9) regardless of the order of the digits.

Exercise 5

Amira wanted to find a 3-digit prime number of the form *aaa*. Is this possible? If so, which primes might she have found?

Solution 5

No. The digit sum of the number is $S(aaa) = a + a + a = 3 \times a$. Such a number is always divisible by 3 regardless of the digits used.

Exercise 6

Write 27 as a sum of distinct primes.

Solution 6

$$2 + 3 + 5 + 17 = 27$$

Exercise 7

Two prime numbers have a sum of 999. Which numbers are they?

Solution 7

All primes are odd, except for 2. Since the sum of two odd numbers is even, 999 cannot be the sum of two odd primes. Therefore, one of the primes must be 2 and the other one must be 997.

Exercise 8

The sum of the first 999 prime numbers is:

(A) odd

(B) even

(C) neither

Solution 8

A list of the first primes must include 2, which is even, while all the other primes are odd.

Since there are 998 odd primes in the sum, the sum of all the odd primes is even. Adding 2 to it will not change its parity.

The correct answer is (B).

Exercise 9

Find three prime numbers that add up to another prime number.

Solution 9

Since 2 is the smallest prime and all the primes larger than it are odd, by adding three primes we must get an odd sum.

The sum can only be odd if the three primes are all odd. Using trial and error, we find several examples:

$$3 + 7 + 13 = 23$$
$$5 + 7 + 11 = 23$$
$$7 + 11 + 13 = 31$$

Have you found any of these or different ones?

Exercise 10

Find the smallest prime number that is the average of two different prime numbers.

Solution 10

$$\frac{3 + 7}{2} = 5$$

The smallest prime that is the average of two different primes is 5.

SOLUTIONS TO MISCELLANEOUS PRACTICE

Exercise 1

Dina has to replace X and Y by two different digits so that the following addition is correct.

$$\begin{array}{r} X\,Y\,X \\ X\,Y\,Y \\ \hline 1\,9\,7\,7 \end{array} +$$

Find the product of the digits X and Y.

Solution 1

To obtain a 4-digit sum, there must be a carryover from adding the two digits denoted by X.

By adding two identical digits, we can only obtain an even sum. Therefore, there must be a carryover from adding the two digits denoted by Y as well as from adding X and Y.

Since the largest sum of two digits is 18, X must be 9. It follows that Y must be 8 and the correct solution is:

$$989 + 988 = 1977$$

The product of the two digits is 72.

Exercise 2

A positive integer is 4 times larger than another integer. Their difference cannot be:

(A) 51

(B) 513

90

(C) 5139

(D) 5191

Solution 2

The difference must be a multiple of 3. Use the digit sum test to find out that 5191 is the only choice that is not divisible by 3. The correct answer is (D).

Exercise 3

Dina, Lila, and Amira pulled three consecutive numbers greater than 10 out of a magical hat. Dina said: "Wow! The digit sum of my number is equal to the sum of the digit sums of your numbers!" Which of the following statements must be true? Check all that apply.

(A) Dina's number is the sum of the numbers held by Lila and Amira.

(B) Dina's number is a repdigit.

(C) Lila's number has the same number of digits as Amira's number.

(D) Any three consecutive numbers have this property.

Solution 3

The digit sums of consecutive numbers are also consecutive numbers, unless there is a change in the number of digits within the sequence. Therefore, Dina's observation can only be true if the numbers are 1, 2, and 3 - impossible since we know the numbers are larger than 10 - or if the sequence of consecutive numbers is similar to one of the following:

$$98, \ 99, \ 100 \qquad S(98) = 17, \quad S(99) = 18, \quad S(100) = 1$$
$$998, \ 999, \ 1000 \qquad S(998) = 26, \quad S(999) = 27, \quad S(1000) = 1$$
$$\cdots \qquad \qquad \cdots$$

Therefore, Dina must hold the middle number which is, in all cases, a repdigit. The correct answer is (B).

The examples show that (A) is not true. However, (A) would be true if we allowed the sequence to be 1, 2, 3.

Since the property is based on a change in the number of digits within the sequence, Lila and Amira must have numbers with different numbers of digits. The choice (C) is invalid.

Finally, not all consecutive numbers have this property. Example: 15, 16, 17 have the digit sums 6, 7, and 8 but $7 \neq 6 + 8$. The choice (D) is invalid.

Exercise 4

Lila, Amira, and Dina each had 99 pennies they wanted to use for their artwork project. Their projects were different, so Lila gave Amira 19 of her pennies, Amira gave Dina 29 of her pennies, and Dina gave Lila 9 of her pennies. Which of the following graphs shows the number of pennies they each had after the exchanges?

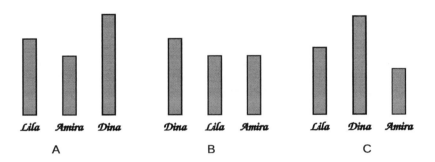

Solution 4

Lila gave away 19 pennies and gained 9. She ended up with 10 pennies fewer than she had at the start.

Amira gave away 29 pennies and gained 19. She also ended up with 10 pennies fewer than she had at the start.

Dina gave away 9 pennies and gained 29. She ended up with 20 pennies more than she had at the start.

Lila and Amira ended up having the same number of pennies. The correct answer is (B).

Exercise 5

The Djinn Jiskan will give Ali and Baba the key to the Sand Treasure if they can point out which of the following statements must be false:

(A) Four consecutive numbers can have a sum that is divisible by 4.

(B) Four consecutive numbers can have a sum that is divisible by 2.

(C) Four consecutive numbers can have a sum that is divisible by 3.

(D) Four consecutive numbers can have a sum that is divisible by 5.

Ali and Baba are treasure hunters, not mathematicians - please help them get the key!

Solution 5

Strategy 1 (concrete): Use easy examples to check each statement. The identity:

$$3 + 4 + 5 + 6 = 18$$

proves that it is possible for 4 consecutive numbers to have a sum that is divisible by 2 as well as by 3.

Also,

$$1 + 2 + 3 + 4 = 10$$

proves that it is possible for 4 consecutive numbers to have a sum divisible by 5 as well as by 2.

By elimination, choice (A) is the correct answer.

Strategy 2 (abstract): Denote the numbers by $N-1, N, N+1, N+2$ and add them:

$$N - 1 + N + N + 1 + N + 2 = 4N + 2 = 2(2N + 1)$$

Since $2N + 1$ is odd for any N, the sum cannot be a multiple of 4. It is clear that the sum is a multiple of 2 for any N. The sum is a multiple of 3 if $N = 4$ and a multiple of 5 if $N = 2$.

Exercise 6

How many pairs of numbers ranging from 20 to 49 are multiples of one another?

Solution 6

40 is a multiple of 20;

42 is a multiple of 21;

. . .

48 is a multiple of 24;

There are 5 pairs of numbers.

Exercise 7

Two positive integers have a product that ends in three zeros and a sum of 261. One of the integers is an odd 3-digit integer, the other one is an even 3-digit integer, and neither of them ends in zero. What is the difference between the two integers?

Solution 7

If the product ends in three zeros, there are at least 3 factors of 2 and at least 3 factors of 5. Since one number is odd and the other is even, all the factors of 2 must be part of the even number. Since neither of the two numbers ends in zero, all the factors of 5 must be part of the odd number.

Therefore, the even number is a 3-digit multiple of 8 and the odd number is a 3-digit multiple of 125. The only multiples of 125 smaller than 261 are 125 and 250. However, if one of the numbers were equal to 250 the other number would not be a 3-digit number. Therefore, the odd number is 125 and the even number is 136 which is, indeed, a multiple of 8.

The difference between the two numbers is $136 - 125 = 11$.

Exercise 8

The Djinn wanted to meet Ali, who was in Cairo, and Baba, who was in Marrakech. The Djinn waited for them in the Seba oasis. The Djinn knew that Ali was always 45 minutes late for appointments while Baba always arrived 35 minutes early. If the Djinn wanted to meet them at exactly 8 AM, what time did he tell each of them to meet him at?

Solution 8

The Djinn asked Ali to arrive at 7:15 AM and Baba to arrive at 8:35 AM.

Exercise 9

Which number does K represent?

$$K + K + K + 333 = 2109$$

Solution 9

Both numbers are divisible by 3. Therefore,

$$
\begin{aligned}
K + 111 &= 703 \\
K &= 592
\end{aligned}
$$

Exercise 10

Dina added 6 to a number and then multiplied the result by 11 to get 1111. What number did Dina start with?

Solution 10

$1111 \div 11 = 101$. Dina started with the number $101 - 6 = 95$.

Exercise 11

The digit product of a number ends in two zeros. The number has at least:

(A) 2 digits

(B) 3 digits

(C) 4 digits

(D) 5 digits

Solution 11

If a product ends in two zeros, then there are at least two factors of 2 and at least two factors of 5 in the product. Digits cannot be larger than 9 - therefore, the two factors of 5 must be separate digits. The factors of 2 can be combined into a digit of 4 (8 also works.)

The number has at least 3 digits. Examples of such numbers: 554, 558, 855, etc.

Exercise 12

A fifth of the students enrolled in Stephan's tennis classes is 4 students less than one third of the enrolled students. If each student has a weekly hour-long lesson, for how many hours does Stephan coach each week?

Solution 12

A third of the students differs from a fifth of the students by two fifteenths:

$$\frac{1}{3} - \frac{1}{5} = \frac{5 - 3}{15} = \frac{2}{15}$$

A diagram with boxes can be used to illustrate this fact:

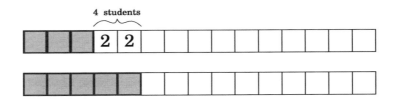

4 students represent two fifteenths of the total number of students. One fifteenth represents 2 students. The number of students is $15 \times 2 = 30$. Stephan coaches for 30 hours each week.

Exercise 13

The sum of a number and its digit sum is 117. How many numbers with this property are there?

(A) 0

(B) 1

(C) 2

(D) 116

Solution 13

The number is smaller than 117, therefore it has 1, 2, or 3 digits. A one digit number added to its digit sum is simply doubled and the result is at most 18. There are no solutions with only 1 digit.

If the number has two digits, the largest digit sum is 18. The solution, if it exists, is a number between $117 - 1$ and $117 - 18$. The only possibility is 99:

$$99 + 9 + 9 = 117$$

If the number has three digits, then the smallest digit sum is 1 and the number is between 100 and 116. The only solution is 108:

$$108 + 1 + 8 = 117$$

The correct answer is (C).

Exercise 14

Jared had 1000 crickets and 7 boxes. He put crickets in each box without counting them. Jared asked several assistants to count the crickets and tell him what they found. After they did so, Jared began to suspect that several assistants did not know much math. Which of the following statements gave Jared this impression?

(A) "Each box contains an odd number of crickets."

(B) "Each box contains the same number of crickets."

(C) "There are 2 boxes with even numbers of crickets and 5 boxes with odd numbers of crickets."

(D) "The numbers of crickets in the boxes form a sequence of consecutive numbers."

Solution 14

Statement (A) is wrong because 1000 is even and the sum of 7 odd numbers must be odd.

Statement (B) is wrong because 1000 is not divisible by 7.

Statement (C) is wrong for the same reason as (A): the parity does not match. 5 odd numbers have an odd sum. (2 odd numbers have an even sum. All pairs of odd numbers add up to an even sum. We add the odd number without a pair to the even sum and the total becomes odd.)

Statement (D) is wrong because the sum of an odd number of consecutive numbers is equal to the product of the middle number and the number of numbers. If the middle number is k, then the sum is $7k$. Since 1000 is not a multiple of 7, it cannot be equal to $7k$ for any k.

Exercise 15

Which number should be removed from the list for the remaining numbers to have a sum that is divisible by 6?

$$61, \ 98, \ 81, \ 114, \ 83, 142$$

Solution 15

Make a list of the remainders that result from dividing each number by 6:

$$1, \ 2, \ 3, \ 2, \ 5, \ 4$$

Group the remainders to form sums of 6: $1 + 2 + 3 = 6$ and $2 + 4 = 6$.

Remove 83 from the list and the sum of the remaining numbers will be divisible by 6.

Exercise 16

What are the next two terms of the following sequence?

$$100, \ 101, \ 110, \ 111, \ 1000, \ 1001, \ 1010, \ 1011, \ 1100, \ \cdots$$

Solution 16

The sequence is formed by all the numbers starting from 100 that can be written using only the digits 1 and 0, in increasing order.

```
    1 0 0
    1 0 1
    1 1 0
    1 1 1
  1 0 0 0
  1 0 0 1
  1 0 1 0
  1 0 1 1
  1 1 0 0
  1 1 0 1
  1 1 1 0
```

Exercise 17

What are the next two terms of the following sequence?

$$N, \; Y, \; YN, \; YY, \; YNN, \; YNY, \; YYN, \; YYY, \; YNNN, \; YNNY, \; \cdots$$

Solution 17

From right to left:

The symbol in the rightmost place changes at each term.

The symbol in the 2nd place changes every 2nd term.

The symbol in the 3rd place changes every 4th term.

```
                N
                Y
          Y     N
          Y     Y
       Y  N     N
       Y  N     Y
       Y  Y     N
       Y  Y     Y
    Y  N  N     N
    Y  N  N     Y
    Y  N  Y     N
    Y  N  Y     Y
```

Exercise 18

Dina and Lila played a number game. One of them thought of a positive integer and the other one asked questions with yes or no answers to find out which number it was. Dina thought of a number and Lila asked the following questions:

1. Is it a 2 digit number?
2. Is it a 1 digit number?
3. Is it even?
4. Is it odd?
5. Is it a multiple of 5?

Lila then figured out which number it was.

1. Which number was it?
2. What were Dina's answers to the questions?
3. Which question could Lila have skipped?

Solution 18

The number was 5.

Dina's answers were:

1. No.
2. Yes.
3. No.
4. Yes.
5. Yes.

Lila could have skipped either question 3 or question 4, since even numbers are not odd and vice-versa.

Exercise 19

Write the next two terms in the following sequence:

$$1, \ 2, \ 6, \ 24, \ \cdots$$

Solution 19

Each term is multiplied by the next consecutive integer to obtain the next term:

$$
\begin{aligned}
1 \times \mathbf{1} \\
1 \times \mathbf{2} \\
2 \times \mathbf{3} &= 6 \\
6 \times \mathbf{4} &= 24 \\
24 \times \mathbf{5} &= 120 \\
120 \times \mathbf{6} &= 720
\end{aligned}
$$

Exercise 20

Alfonso received fresh pastry from Max, his baker. There were as many croissants as Danishes, and as many strudels as baklavas. There were four times as many croissants as strudels. The pastries came in boxes of 15 pieces each. Alfonso received 11 boxes, but one of them was not full. How many baklavas did he receive?

Solution 20

Since there are four times as many croissants as baklavas, the sum of the baklavas and the croissants must be divisible by 5. Similarly, the sum of the strudels and Danishes must also be divisible by 5. Since the two sums are equal, the total number of pieces must be divisible by 10.

11 full boxes contain 165 pieces and 10 full boxes contain 150 pieces. Therefore, the total number of pieces of pastry must be larger than 150 and smaller than 165. In this range, only 160 is divisible by 10. Max received 160 pieces of pastry.

Half of these pieces (80) are baklavas and croissants. Of these, one fifth (16) are baklavas.

Exercise 21

Find every two digit number that decreases by 5 when its digits are reversed.

Solution 21

This can only happen if the difference between the two numbers is a multiple of 5. The difference is a multiple of 5 if its last digit is either 0 or 5. Thus, the difference of the two available digits must be equal to either 5 or 0. If it is zero, then both numbers are zero and cannot differ by 5. Therefore, the difference between the two digits must be 5.

There are 5 pairs of digits that satisfy this requirement:

$$(5, 0), \ (6, 1), \ (7, 2), \ (8, 3), \ (9, 4)$$

Since the pair $(5, 0)$ does not yield a 2-digit number when reversed, there are 4 pairs left. The numbers are:

$$61, \ 72, \ 83, \text{ and } 94$$

Exercise 22

In the following calendar, the dates hidden by the letters Q and S could have the same sum as the dates hidden by:

Sun	Mon	Tue	Wed	Thu	Fri	Sat
		P	W	Q		
		S	T	V		
		R	M	K		

(A) W and T

(B) V and R

(C) R and W

(D) M and P

Solution 22

The correct answer is (A).

(A) W is smaller than Q by 1 and T is larger than S by 1, therefore the two pairs of dates have the same sum.

(B) V is larger than S by 2 and R is larger than Q by 12, therefore the dates hidden by them have a sum larger that the desired sum by 14.

(C) R is larger than S by 7 and W is smaller than Q by 1, therefore the dates hidden by them have a sum larger than the desired sum by 6.

(D) M is larger than S by 8 and P is smaller than Q by 2, therefore the dates hidden by them also have a sum larger than the desired sum by 6.

Exercise 23

Alfonso has 14 boxes of onions lined up on a shelf of his store. The number of onions in each box differs from the number of onions in a neighboring box by either 1, 3, 7, or 13. The total number of onions in the 14 boxes cannot be:

(A) 47

(B) 128

(C) 113

(D) 59

Solution 23

Adding or subtracting an odd number to a number, changes the parity since:

$$\text{odd} + \text{odd} = \text{even}$$
$$\text{even} + \text{odd} = \text{odd}$$

This means that the number in each box has different parity from the number in the neighboring boxes. If we group the boxes into pairs of neighboring boxes (7 pairs), each pair must have an odd sum. Since there are 7 odd numbers to add, the result must be odd. The total number of onions is odd.

The answer is (B).

Exercise 24

A 3-digit number is added to its digit sum and the result is 513. What is the number? How many solutions are there?

Solution 24

Assume the number has the digits a, b, and c. Write it in expanded form:

$$abc = 100 \times a + 10 \times b + c$$

and add it to its digit sum:

$$abc + S(abc) = 100 \times a + 10 \times b + c + a + b + c$$

Then:
$$513 = 101 \times a + 11 \times b + 2 \times c$$

Since a, b, and c are digits, they can only have small values. There is no value of b that could make $11 \times b$ exceed 100. Therefore $a = 5$ or $a = 4$.

Case 1: $(a = 4)$

$$
\begin{aligned}
513 &= 404 + 11 \times b + 2 \times c \\
109 &= 11 \times b + 2 \times c
\end{aligned}
$$

Similarly, $2 \times c$ cannot exceed 18 which means that $11 \times b$ cannot be smaller than 91. This means that $b = 9$ and $c = 5$.

The number is 495.

Case 2: $(a = 5)$

$$
\begin{aligned}
513 &= 505 + 11 \times b + 2 \times c \\
8 &= 11 \times b + 2 \times c
\end{aligned}
$$

In this case, $b = 0$ and $c = 4$.

The number is 504.

There are two solutions: 495 and 504.

Exercise 25

The sum of 6 distinct positive integers is 19. The largest number is:

(A) 6

(B) 8

(C) 9

(D) 10

(E) there is no solution

Solution 25

19 is too small to be the sum of 6 different positive integers. The smallest sum that can be obtained with 6 different positive integers is:

$$1 + 2 + 3 + 4 + 5 + 6 = \frac{6 \times 7}{2} = \frac{42}{2} = 21$$

The answer is (E).

Exercise 26

The product of two numbers is 203. If we subtract a number from one of the two numbers, their product becomes 175. Which number was subtracted?

Solution 26

Factor 203 and 175 into primes:

$$203 = 7 \times 29$$
$$175 = 5 \times 5 \times 7$$

Therefore, one of the numbers must be 7. The number that decreased was 29 and it decreased by 4.

Exercise 27

Two of four consecutive numbers have a sum of 21. The sum of the other two cannot be:

(A) 17

(B) 21

(C) 23

(D) 25

Solution 27

Two of four consecutive numbers differ by at most 3. The two numbers that have an odd sum cannot differ by 2 since, in that case, they would be both even or both odd and their sum would be even. This leaves the cases where the two numbers differ by 1 or by 3. The possiblilities are:

$$9, \ 10, \ 11, \ 12$$
$$10, \ 11, \ 12, \ 13$$
$$8, \ 9, \ 10, \ 11$$

In the first case the numbers chosen can be either 9 and 12 or 10 and 11. In either case, both sums are 21.

In the second and third cases, the sums of the numbers that were not chosen are $8 + 9 = 17$ and $12 + 13 = 25$.

The sum 23 cannot be obtained. The correct answer is (C).

Exercise 28

Lila had a paper rectangle with length 16 units and width 12 units. She repeatedly folded it in half until she could no longer obtain integer length sides. How many folds did she make?

Solution 28

Each time she folded the paper in half, one of the dimensions - either the length or the width - got halved. The process stopped when both dimensions became odd. Factor the dimensions into primes, to see how many factors of 2 there are. There will be as many folds.

$$16 \ = \ 2 \times 2 \times 2 \times 2$$
$$12 \ = \ 2 \times 2 \times 3$$

There are 6 factors of 2, therefore there will be 6 folds.

Note: there are multiple ways to fold the paper.

Competitive Mathematics Series for Gifted Students

Practice Counting (ages 7 to 9)
Practice Logic and Observation (ages 7 to 9)
Practice Arithmetic (ages 7 to 9)
Practice Operations (ages 7 to 9)

Practice Word Problems (ages 9 to 11)
Practice Combinatorics (ages 9 to 11)
Practice Arithmetic(ages 9 to 11)
Practice Operations (ages 9 to 11)

Practice Word Problems (ages 11 to 13)
Practice Combinatorics (ages 11 to 13)
Practice Arithmetic and Number Theory (ages 11 to 13)
Practice Algebra and Operations (ages 11 to 13)
Practice Geometry (ages 11 to 13)

Practice Word Problems (ages 12 to 15)
Practice Algebra and Operations (ages 12 to 15)
Practice Geometry (ages 12 to 15)
Practice Number Theory (ages 12 to 15)
Practice Combinatorics and Probability (ages 12 to 15)

This is a series of practice books. With the exception of a few reminders, there are no theoretical explanations. For lessons, please see the resources indicated below:

Find a set of free lessons in competitive mathematics at www.mathinee.com. Addressing grades 5 through 11, the *Math Essentials* on www.mathinee.com present important concepts in a clear and concise manner and provide tips on their application. The site also hosts over 400 original problems with full solutions for various levels. Selectors enable the user to sort essentials and problems by test or contest targeted as well as by topic and by the earliest grade level they can be used for.

Online problem solving seminars are available at www.goodsofthemind.com. If you found this booklet useful, you will enjoy the live problem solving seminars.

For supplementary assessment material, look up our problem books in test format. The "Practice Tests in Math Kangaroo Style" are fun to use and have a well organized workflow.

Made in the USA
San Bernardino, CA
07 May 2016